SAE EDGE™ RESEARCH REPORT

# Unsettled Domains Concerning Electric Propulsion Technology for Commercial Aircraft

**Ravi Rajamani, PhD**
***Principal Consultant, drR2 consulting, Connecticut, USA***

EDGE DEVELOPMENT TEAM

Richard Ambroise, *Airbus*
Rodger Dyson, *NASA Glenn Research Center*
Craig Lawson, *Cranfield University*
Steve O'Flarity, *Pratt & Whitney*
Norm Pereira, *FAA*
Joshua Portlock, *Electro.Aero*
Gilles Rosenberger, *Faraday Aerospace*
Michael Winter, *Pratt & Whitney*

SAE INTERNATIONAL®
Warrendale, Pennsylvania, USA

## About the Publisher

SAE International® is a global association of more than 128,000 engineers and related technical experts in the aerospace, automotive, and commercial vehicle industries. Our core competencies are life-long learning and voluntary consensus standards development. Visit sae.org

## SAE EDGE™ Research Report Disclaimer

SAE EDGE™ Research Reports focus on topics that are dynamic, in which knowledge is incomplete, and which have yet to be standardized. They represent the collective wisdom of a group of experts and serve as a practical guide to the reader in understanding unsettled subject matter. They are not meant to provide a recommended practice or protocol. The experts engaged have contributed their own thoughts and points of view, and these are not the positions of the institutions or businesses with which they are affiliated. A professional writer has collectivized their input; there is no one contributor's perspective being advanced, but rather that of a community of practitioners. SAE EDGE™ Research Reports are the property of SAE International and SAE alone is responsible for their content.

## About This Publication

SAE EDGE™ Research Reports provide state-of-the-art and state-of-the-industry examinations of the most significant topics in mobility engineering. SAE EDGE™ contributors are experts from research, academia, and industry who have come together to explore and define the most critical advancements, challenges, and future directions in areas such as vehicle automation, unmanned aircraft, IoT and connectivity, cybersecurity, advanced propulsion, and advanced manufacturing.

## Related Resources

**Fundamentals of Electric Aircraft**
**Electric Flight Technology: The Unfolding of a New Future**

## SAE Team

Frank Menchaca, Chief Product Officer
Michael Thompson, Director, Standards, Information and Research Publications
Monica Nogueira, Acquisitions Director
Jill Leonard, Product Manager
William Kucinski, Managing Technical Editor



Printed in USA



**ISSN 2640-3536**
**e-ISSN 2640-3544**
**ISBN 978-1-4686-0115-2**

**To purchase bulk quantities, please contact:** SAE Customer Service

E-mail: CustomerService@sae.org
Phone: 877-606-7323 (*inside USA and Canada*)
724-776-4970 (*outside USA*)
Fax: 724-776-0790

**Visit the SAE International Bookstore at books.sae.org**

## About the Editor

**Dr. Ravi Rajamani** is an independent consultant in the aerospace and energy sectors with many years of experience in the application of systems engineering principles, data analytics, and model-based methods to controls, diagnostics, and prognostics, especially for propulsion systems. He has three books to his name including *Electric Flight Technology: The Unfolding of a New Future*. In addition, Dr. Rajamani is the author of many book chapters, journal and conference papers, and has several patents. Prior to his current job, Ravi worked at Meggitt, United Technologies Corporation, and the General Electric Company. He has a PhD from University of Minnesota and an MBA from University of Connecticut. His earlier degrees were a BTech from IIT, Delhi, and an MSc from IISc, Bangalore. He is active within various SAE technical committees dealing with prognostic health management (PHM) and electric propulsion. He is also active in the PHM Society, serving on its board of directors. He has been elected a fellow of SAE International and of IMechE. He currently serves as the editor in chief of the *SAE International Journal of Aerospace*. In addition, he has a research appointment at the University of Connecticut and is a visiting professor at Cranfield University.

# contents

Atosan/Shutterstock.com

# Unsettled Domains Concerning Electric Propulsion Technology for Commercial Aircraft


## Abstract

According to some, electric propulsion is already on its way down the "trough of disillusionment." This report argues that while there are some concerns with associated technologies, such pessimism is unwarranted. Yes, battery systems can be more energy dense; charging standards are late in being developed; it is not clear what technologies and architectures will win out for which sectors; we still have to figure out how to deal with thermal management issues at all levels in future designs; and what might be the certification standards for electric aircraft. But that does not mean that these issues will not be addressed. Yes, these unsettled questions may take a little longer to solve than originally estimated, but there is full expectation within the industry that electric propulsion for commercial aircraft will succeed. In this SAE EDGE Research Report we present points of view from leading researchers in the industry who are thinking deeply about solving these problems.

NOTE: SAE EDGE™ Research Reports are intended to identify and illuminate key issues in emerging, but still unsettled, technologies of interest to the mobility industry. The goal of SAE EDGE™ Research Reports is to stimulate discussion and work in the hope of promoting and speeding resolution of identified issues. SAE EDGE™ Research Reports are not intended to resolve the issues they identify or close any topic to further scrutiny.



**RAVI RAJAMANI, PhD**
***Principal Consultant,***
***drR2 consulting, Connecticut, USA***

### Edge Development Team

Richard Ambroise, *Airbus*
Rodger Dyson,
*NASA Glenn Research Center*
Craig Lawson, *Cranfield University*
Steve O'Flarity, *Pratt & Whitney*
Norm Pereira, *FAA*
Joshua Portlock, *Electro.Aero*
Gilles Rosenberger, *Faraday Aerospace*
Michael Winter, *Pratt & Whitney*



ISSN 2640-3536

# Introduction

## State of the Industry

So, where are we in the "hype cycle" for aircraft electric propulsion? Early in 2019, a friend shared an article [10] that reported on the moderating views expressed by the chief technology officer (CTO) of Airbus about the feasibility of an all-electric (AE) aircraft and speculated that we may have started our descent into the "trough of disillusionment!" While many claims made in that article are accurate, the writer does not consider all the reasons why electric aircraft (EA) would find purchase among aviation experts despite the negatives. In this SAE EDGE Research Report, we will try to give a balanced view of where EA technology is headed, what are some of the fundamental issues with it, and how are technologists trying to overcome them. We will also report on some of the predictions that leading researchers in the field are making, and why there is merit in their views.

Electrification has been called the third revolution in aviation. This follows the birth of heavier-than-air machines with the Wright brothers in 1903, which constituted the first revolution, and the advent of jet propulsion during and after World War II, which some consider the second revolution. It is interesting that lighter-than-air machines are not given much importance in this chronology, even though the first attempts at electric propulsion were carried out on such airships.

At any rate, the current frenzy of research and development all around the world in EA surely defines a radically different era in aviation; one that we have not seen in a long time. According to Robert Thomson of Roland Berger, there are more than 170 EAs being developed around the world, and this number would probably exceed 200 by the end of 2019 [27]. This has not prevented some spectacular and high-visibility failures in the field as well. The latest is the shutdown of Zunum Aero, after Boeing pulled its support [5]. In June 2019, Boeing announced a strategic partnership with Kitty Hawk to further develop the Cora. This partnership will definitely help with issues like certification, where Boeing has several decades of experience.

Clearly, nothing is a given in this fast-moving industry. If all its touted benefits are realized, electric propulsion will completely change the face of commercial aviation. It will also take this sector out of the bull's-eye of environmentalists, who declaim its deleterious effects on our climate. Even though there is a target on its back because of the sheer amount of fuel burned and the greenhouse gases (GHG) emitted per passenger, in absolute terms, the aviation sector is not that bad an actor. The entire aviation industry accounts for only about 2% to 4% of the environmental impact caused by emissions from all sources, even though, as we noted, on an individual basis it is pretty bad. A European Environmental Agency 2014 report comparing different forms of transportation modalities estimated that aircraft emit nearly 2000% more carbon dioxide than a train when compared on a passenger-mile basis. Because of its huge relative impact on the environment, any work on reducing aviation-related emissions is admirable; however, it should also be understood that because of its low absolute impact, electrification of aviation will make little difference to overall GHG concentrations in the atmosphere. But, as other sources of GHG are cleaned up, aviation will become a bigger and bigger piece of the puzzle. The focus will naturally shift and the pressure to electrify will grow stronger. While technological challenges to electrification will remain, at least in one aspect things will become easier. The wholesale electrification of ground vehicles will result in the entire transportation infrastructure shifting to support electric propulsion, positively affecting the aviation sector as well.

In this report we will tackle a few of the more unsettled topics in electric propulsion in the aviation sector. One, of course, must be very careful when defining the boundaries of any such discussion. The argument for one specific kind of aircraft will not apply to another, and therefore, qualifications are important. AE aircraft will have different technological challenges and characteristics when compared to hybrid-electric (HE) and more-electric (ME) architectures. To clarify some of the terms in this field, we will use the following definitions for the different kinds of EA architectures: AE, HE-parallel (HEP), and HE-series (HES). There are many variations and refinements of these within the EA literature, but ultimately, they all reduce to one of these three forms.

In the AE architecture the source of electric energy is a solar panel or a battery. In this report we are not considering solar technology. Solar flight has had an eventful history since about the 1970s (see the books by Rajamani [19] or Pascal Thalin [26] for details), and even accomplished a circumnavigation of the globe in 2015-16. But as a practical means for transporting an economically viable payload it is absurdly inadequate. A quick calculation shows that even to carry one person, a solar airplane needs a wingspan rivaling that of a conventional twin-aisle jet [26]. This makes solar power infeasible for commercial aviation applications. However, solar energy has an important role to play in light-payload applications such as communications, where a well-designed aircraft can stay aloft indefinitely with zero additional fuel and close-to-zero environmental impact. But this is not the forum to discuss that.

In Figure 1, all the green propulsors are electric fans and the red propulsors are turbines. HE systems depend on hydrocarbon fuel-burning engines to generate the electrical energy to drive the fans. While in a parallel system (HEP) the engines provide propulsive power in addition to electric energy, in the series system they only drive the generator that produces the electricity. With the HES architecture, having a storage system like a battery would help even out the supply and the demand with respect to electric power. With the right-sized propulsors and generation system, this is not an absolute necessity.

Before we go farther, we need to deal with a system that has always been posited as a viable alternative to batteries or generators for providing electric energy: the fuel cell. How does one characterize an aircraft using a fuel cell as

**FIGURE 1.** Different EA architectures.

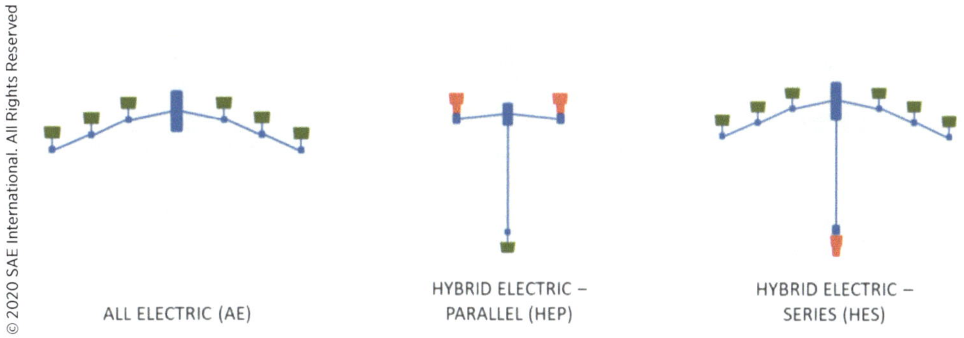


its primary source of energy? While this does not have an engine-generator combination, a fuel cell nevertheless converts a fuel (either hydrogen or a hydrocarbon) into electricity, so it can more correctly be classified as an HES design rather than an HEP design. There are a few outfits around the world that are working on hydrogen-fueled aircraft, most of them retrofitting existing conventional platforms with fuel cells. A German consortium (H2Fly) has modified a Pipistrel 4-seater, and a company in California-ZeroAvia-plans to modify a Piper aircraft. Boeing was probably the first company to fly a fuel cell-powered aircraft when its Madrid research team division modified a Diamond motor glider with a proton-exchange membrane fuel cell back in 2008.

But in the larger scheme of things, this is a relatively niche application, and for this report we are going to stick to conventional engines. Fuel cells will not significantly change the economics of electric propulsion in the foreseeable future. Further down the road, however, there could be a role for them. Hydrogen happens to be the most energy-dense compound in the world, not counting fissionable material. Liquid hydrogen is also cold enough to sustain cryogenic temperatures. These two properties can be used very effectively in a future advanced electric system, where the cold liquid form keeps electrical components at cryogenic temperatures, and gaseous hydrogen can be used in a combustion system or a fuel cell to generate electric energy. The National Aeronautics and Space Administration (NASA) is studying novel methodologies incorporating these ideas and may one day discover unique solutions that would solve many of the outstanding problems of electric propulsion. However, that day is not today; for now we are still working with more conventional methodologies.

The question of environmental impact was already mentioned above. A related topic is the use of energy and the overall efficiencies associated with electrification. Ground vehicles are a little different in this regard because of their usage patterns. Consider the plug-in electric-hybrid vehicles (PEHV) as an example. They go through several start-stop cycles during the course of a trip, especially if that trip is in an urban setting. For ground vehicles, most of the journeys actually are of this kind. Governments are aggressively supporting their electrification with subsidies and, in some cases, outright edicts that will ban hydrocarbon-powered vehicles after certain dates. There is help being provided for the installation of a charging infrastructure and a general awareness of the need to reduce the use of GHG-emitting vehicles. Because of these constraints and requirements, electric vehicles are designed with different considerations than are electric aircraft. This results in greater reductions in emissions and an increase in efficiency when compared to corresponding figures for aerospace vehicles.

Management of heat is a critical aspect of EA. In conventional aircraft, several factors help with the dissipation of the heat of combustion, the chief being the "dumping" of the hot exhaust gases overboard. Other locations can be cooled with the use of heat exchangers that use the free flow of outside air to their advantage. With a electrical system in an AE aircraft, this simple solution is not available. The heat generated by the power electronics and other electrical components needs to be dissipated in ways other than being dumped overboard. The use of different kinds of heat exchangers has been considered, as has been the use of cryogenic systems to reduce the generation of heat during operation, but all of these have many cons. The bottom line is that this is a real problem that needs to be dealt with during design, and it is not clear whether there are viable solutions for larger aircraft where the heat burden can become prohibitive.

Battery and battery systems are central to many configurations of EA and bring up a number of unsettled issues. The biggest problem with batteries is that they are extremely heavy for the amount to energy they store and deliver. Secondary, or rechargeable, batteries are even worse than primary, or non-rechargeable, batteries; probably four times worse in energy density. This means that, compared to internal combustion reciprocating engines or gas turbines, their power-to-weight ratio is extremely low, which seriously affects its range and efficacy as a means of propulsion. A related issue with batteries used in an AE aircraft architecture is that they need to be plugged in and charged before use. Given the muddled history of charging systems for electric vehicles, it would be nice if the same or similar problems could be avoided by the early adoption of global charging standards. This is not only a question of form and fit, but it also refers to the voltage, current, and power levels to cover the range of aircraft that might need to be plug-charged. Related issues such as communications and cooling should also be considered.

Another issue with batteries, as compared to fuel-burning engines, is that the conventional aircraft uses up fuel during the course of the flight. Hence, as an example, the demands on the landing gear during landing are much less than if it were to land with a full tank of gas. In fact, aircrafts have to dump fuel before undertaking an emergency landing just after takeoff. With AE aircraft, there is no reduction in the weight of the aircraft as it uses up its stored energy, so there is no relief for the landing gear designer, which means extra weight and structural integrity for achieving comparable operational life.

There are other issues with electric propulsion that we will not deal with in detail but mention in passing. Issues such as electric actuation, ice protection, accident response, system certification, financing, and so on. Some of these are technical in nature, but many others are related more to the business or social aspects of the aviation industry.

**A Brief History** It is not our intention in this report to give a detailed history of this industry. For that, the reader can consult any number of recent books, including one by this author [19] and the one by Thalin et al. [26]. It is given here just to set this work in the right historic context.

As mentioned above, the first experiments with electric propulsion were by French balloonists at the end of the nineteenth century. The Tissandier brothers have been credited with being the first to successfully demonstrate this in 1883. Figure 2 shows their airship, which was fitted with a Siemens motor powering a propeller. The source for the electric energy seems to have been a zinc accumulator (battery) of their own design.

It was not until nearly a century later, in 1973, that an aircraft (heavier than air, at this time) flew for the first time powered solely by batteries and an electric motor. This was near Linz, Austria, and it was an HB3 Brditschka motor glider modified by Fred Militky for electric power and flown by Heino Brditschka. In the MB-E1, the petrol engine was replaced with a Bosch motor, powered by Varta Ni-Cad batteries. Before this achievement, there were some interesting attempts in the intervening years to exploit electricity to power flight. For example, during World War I, there were attempts to power a helicopter as a reconnaissance platform using a tether. But because the motors were too heavy, this was not very successful. The MB-E1 was the first real EA in the modern sense of the word.

However, even this small experiment did not persevere. After the Austrian achievement, electric propulsion took a back seat to solar flight for the next three decades or so. Solar-powered aircraft started with small remotely controlled unmanned planes built by Roland Boucher's AstroFlight in the early 1970s to Paul MacCready's Solar Challenger that crossed the English Channel in 1979 with a pilot on board. The apotheosis of the solar era came in 2015-2016 when the Swiss pair of Bourschberg and Piccard circumnavigated the globe in their Solar Impulse 2. Unlike the Challenger, the Impulse had batteries, which allowed it to fly during the night as well. While solar flight will always have a place in aviation for niche applications, such as communications or

**FIGURE 2.** **Tissandier Balloon near Auteuil, France, in 1883.**

From Tilly [29]

reconnaissance, it is naïve to expect it to provide any support for manned flight of any magnitude other than for sport.

The modern era of electric flight started more gradually as existing aircraft transitioned to ME operation. Conventional systems such as brakes or some actuation systems evolved from being purely hydraulic systems to being either hybrid systems, using electromechanical and hydraulic elements, or being purely electromechanical systems. In the commercial world, the Boeing B787 and the Airbus A380 are examples of such an ME architecture, and in the military world, the Lockheed Martin F-35 Joint Strike Fighter exploited electrical elements wherever possible. Another impetus for electric flight was the explosion in the development of electric ground vehicles and consequently, the availability of cost-effective and optimized solutions for all aspects of the electric transmission chain. It is important to note that while both domains share many common technologies, there are crucial differences as well. For example, as we will argue further down in this report, the efficiency arguments that apply to the ground-based system do not apply as readily in the air, because the modes of operation of the two vehicles are so different. But nevertheless, the EA industry can thank the electric vehicle industry for many of the technologies it relies on.

Many large, specialized agencies like NASA in the United States and DLR in Germany, and many companies in the United States and around the world took up the challenge of developing electrically powered aircraft or constituent components. The common goal was to help make electric flight technically and economically more feasible. Even large original equipment manufacturers (OEMs), such as Airbus, got into the act, often funded by government research grants. Mitigating the environmental impact of aviation was one of the reasons for these developments. We will see that while this is admirable, in the larger scheme of things, this is probably not that beneficial.

The first attempts at electrification consisted of converting motor gliders by replacing existing engines with an electrical system-much like the MB-E1. Airbus was one of the first companies to design and build an AE aircraft from the ground up. This was the E-Fan that was built by a team lead by Airbus and included Siemens as the electric equipment supplier. Airbus shut down this concern-called VoltAir-in 2017 and is pursuing other ventures, including their electrical vertical takeoff and landing (eVTOL) craft called Vahana being designed in Silicon Valley. In fact, much of the new energy in EA innovations has moved to the "Silicon Valleys" of the world, with people like Larry Page of Google, and companies like Uber, jumping into the electrification game. However, traditional OEMs are not giving up, and most of them have initiated some EA program or the other, either on their own, or in partnership with others. Bell, Boeing, UTC, Leonardo, Honeywell, Snecma, Siemens, Rolls-Royce, Airbus, etc., are all playing the EA game. Some of them are acquiring others to gain a quick foothold; Boeing, for example, acquired Aurora Flight Sciences and Rolls-Royce bought the EA division of Siemens.

Urban air mobility (UAM) or the concept of air taxis is motivating some of this activity. Uber Elevate is making a lot of waves with their stated goal of adding air-taxi services to their ground transportation sector and their intentions of testing these systems in large metropolises around the world. They have partnered with many aerospace companies to futher their UAM concepts, and recently announced another one with a car company, Hyundai, as well. Dubai has been working on a commercial air-taxi system for a while now, with the latest partner being Volocopter. Kitty Hawk, which has now merged with Zee Aero, and is supported by Larry Page, is developing and testing an eVTOL aircraft-Cora-for certification in New Zealand, according to eVTOL.news [9]. There is, clearly, a lot of energy in this sector. Robert Thomson from Roland Berger reported in May 2019, that more than 170 electrically powered aircraft are being built around the world and that this may grow to 200 by the end of 2019 [27]. Whether this happens or not is only mildly interesting. The truth of the matter is that while this is a huge technological growth area, the regulatory and economic pressures surrounding it will mean that nothing is going to move very rapidly; the game will play itself out over the course of the next 10 to 25 years, and we will not really know what the final EA reality is for at least that long. Some are more optimistic than others. Hybrid systems, and especially those converted from existing aircraft, might have a better chance of entering service in the near term, compared to new designs. Ampaire, for example, has converted a Cessna 337 6-seater to a hybrid configuration and has one conventional engine and another electric propulsor. Their path to certification would be through a supplemental type certificate (STC) for the electrical additions. Figure 3 depicts the projected systems that could come in play during the next 30 years or so, as envisioned by ATI in the United Kingdom, which was shared by Professor Lawson at Cranfield.

## Unsettled Domains Concerning Electric Propulsion Technology

There are several domains associated with electric propulsion that are in flux and need to be examined. Some of these are quite fundamental in nature and can affect the way the industry evolves. Others are not necessarily technically unsettled but need the various stakeholders to come together and agree on standards. This way an industry-wide consensus can be reached which would help move it forward in the right direction. In this report, we will examine a few such issues; and as this sector grows and matures, we might examine more in the future. It has never been very clear why there has been such a rapid growth in interest in electric propulsion. Ostensibly, it is because electric power will help to eliminate the environmental impact of aviation. Closely coupled to this is the question of efficiency, because an argument has always been that electrical systems are far more efficient at all speeds when compared to the conventional propulsive system i.e., a

**FIGURE 3.** Electric power demands of future EA platforms.

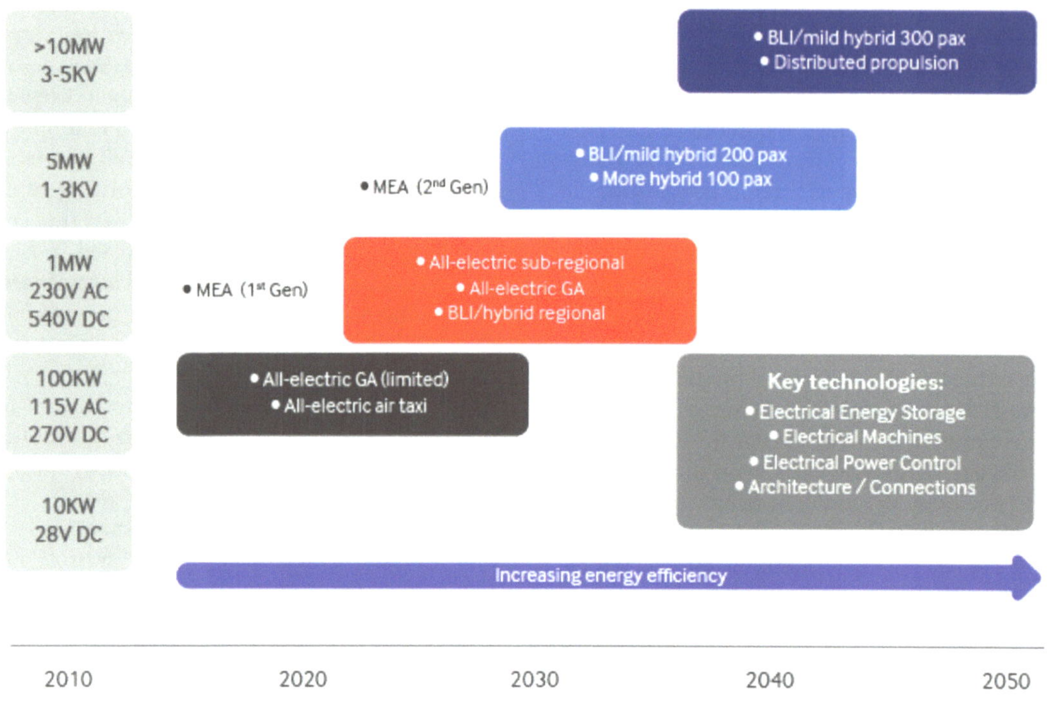




gas turbine (GT) or a reciprocating engine. We will examine the open questions surrounding these claims.

Other open questions revolve around batteries and the systems that enable them. For certain classes of EA (i.e., those that will be powered solely by electric energy) the performance of the battery system determines everything: the range, the safety risk, and the economics. Related to this is the fact that such plug-in vehicles need to be charged, which means that standards need to be developed around these devices. While plugging in the vehicle is one way of charging the aircraft, another way to accomplish this is by developing battery systems that can be easly swapped. It is entirely possible that the industry will evolve in that direction. These two topics will be examined in this report.

Unlike conventional engines, electric motors, and their accompanying drives and controllers, have some disadvantages with respect to the ability to manage thermal issues. This is not a trivial task, as we shall see. Much has been made of superconductivity as well, and we will see whether, and where, these systems make sense.

Finally, we will look at a few system-level issues such as system certification and financing. To summarize, the topics that will be discussed in this SAE EDGE Research Report are:

- The energy and environmental conundrum
- Battery systems
- Thermal management
- Other issues at the system level

## The Energy and Environmental Conundrum

The question that is often asked is, what is the compelling case for electric propulsion? We have delved a little into the fact that while the aerospace industry can be quite polluting on a per-occupant basis, the total GHG emissions are a small part of the overall emissions by all sources around the globe. This means that improving its performance in this area will not do much for climate change.

But what about energy? Today much of the aviation industry is propelled by GTs burning jet fuel. Gasoline is used in reciprocating engines for smaller general aviation (GA) aircraft, but it constitutes a far smaller amount. According to the U.S. Energy Information Administration, the total amount of jet fuel consumed in the United States in 2018 averaged to about 1.7 million barrels per day. The total consumption for the entire transportation sector was 14.2 million barrels. This means that aviation use of hydrocarbons was about 12% of the total transportation use. This number, and also the numbers associated with the environmental impact of aviation that we saw earlier, should be kept in mind when we discuss electrification of the aviation sector. Suffice it so say that going after 12% will not have a dramatic effect on the overall emissions. But there is still an incentive to do this, again mainly because other transportation sectors are moving toward lowering their environmental impact, and the aviation sector does not want to be left behind.

Given that electrification is here to stay, in one form or another, let us examine the figures associated with it. Much of this section is based on discussions with Steve O'Flarity from Pratt & Whitney and his paper (with Alan Epstein) in the AIAA *Journal of Propulsion and Power* [8]. They lay out a very compelling case for why it will be a long time before the benefits of electrification will be realized. In addition, papers by Rodger Dyson from NASA GRC in Cleveland and his collaborators inform some of the views expressed here, as do discussions with Professor Craig Lawson from Cranfield University in the United Kingdom.

## Environmental Impact

A criticism that has been levied against electric vehicles is that they are not as environmentally friendly as advertised. This is because, according to the critics, the emissions associated with the manufacture and transportation of the vehicle and its components, such as the battery, are not taken into account. This is a valid argument, even if the conclusions reached are not completely accurate. The Union of Concerned Scientists (USC) did an extensive study, reported in 2012, that examined all aspects of an electric vehicle and compared them with conventional vehicles. *The New York Times* reported [23] at that time that the amount of well-to-wheel emissions from an electric vehicle depends drastically on where it is operated, and hence how green is the electricity it is using to power up. This is obvious but not something that is normally discussed. Also analyzed was the amount of environmental impact the manufacturing processes had. Building an electric vehicle does generate more GHG-even 68% larger for the largest cars-than does the manufacture of conventional vehicles. But even when all this was factored in, electric vehicles came out way ahead-for some cars twice as green as compared to equivalent gas-powered ones-except against the most efficient conventional vehicles operating in the areas with the dirtiest electricity. That is why many governments are moving ahead with mandating the exclusive use of electric vehicles within the next 10 to 20 years. This criticism, based on the well-to-wheel numbers, applies to EA as well. We shall examine some of these claims in this section.

According to the International Civil Aviation Organization (ICAO), a UN specialized agency, the total $CO_2$ emissions by the entire aviation sector amounted to about 2% of the emissions from all sources [12]. Because emissions of other harmful gases from aircraft occur at higher altitudes than from other emitters, it is further estimated that the deleterious effect of GHG emissions from flying aircraft is disproportionately larger, possibly up to twice as much. Regardless of this, the bottom line is that the environmental impact of all of aviation is a tiny portion of the total from all anthropogenic (i.e., resulting from human activity) causes. For comparison (see Figure 4), considering just the industrial sector, 44% of $CO_2$ emissions come from power generation; road transportation accounts for 17%; buildings another 10%; and shipping 3% (according to the Air Transport Action Group, admittedly a partisan advocacy organization, quoting myriad other scientific sources [3]). Decidedly, if the aviation sector is left unchecked while all other sectors are more strictly regulated, this 2% number will grow bigger, so this is not a call for being pollyannaish about our industry. It is, however, a plea to consider the facts and to be realistic about the current situation and about the prospects of electrification. There is still a lot of "runway" left in conventional GT technology to reduce GHG further and, given all the issues with electrification, these should probably be exploited first before we make a major shift to an electric future.

While aviation is not a huge contributor to global warming, it is nevertheless true that the airline industry is in a healthy growth phase. Passenger numbers are growing by about 5% per year, and they will grow even more dramatically as more and more newly prosperous Chinese and Indians

**FIGURE 4.** **The relative and the absolute impact of transportation on the environment.**

Relative environmental impact in industrial sector

| Source | Percent |
|---|---|
| Power generation | 44% |
| Road transportation | 17% |
| Buildings | 10% |
| Ships | 3% |
| Aviation | 2% |

Source: Air Transport Action Group

Absolute environmental impact of different forms

| Source | $CO_2$ (g) |
|---|---|
| Train | 14 |
| Bus | 68 |
| Motorbike/Scooter | 72 |
| Car | 73 |
| Plane | 285 |

Quoted per passenger km

Source: European Environmental Agency & Wikipedia

take to the skies. So, everything must be done to try and curb the increase in aviation's impact on the environment. But is electrification the way to go? This is very much still unsettled.

Electricity for propulsion has to come from somewhere, either generated by the aircraft itself in an HE system or from the electric outlet on the ground in an AE system. Regardless, when calculating the environmental impact of the aircraft, one has to take into consideration the entire value chain associated with the system. At the very least, one must consider the GHG emissions caused by the producers and distributors of the energy and the battery, with due consideration given to the fact that batteries have lives that are much shorter than the airframe itself. This means that the battery stacks-consisting of several cells-would be replaced many times during the life of the aircraft, and therefore, the environmental impact of the manufacturing process cannot be discounted. Also, much of the world's electricity is still generated by coal plants, at least in two countries where EA are seriously being considered for commercial operation: the United States and China. And these plants, even with advanced "clean-coal" technologies, are quite polluting when compared to natural gas or nuclear energy plants. In fact, it is because of this that the average carbon dioxide emissions of the world power plants today falls somewhere in between that from aircraft sources (Figure 5). Until the world switches substantially to renewables, this problem will persist and must be accounted for. Given this, even an AE aircraft will have some impact on the environment, albeit not as much as a fuel-burning conventional engine, and definitely at a diminishing rate as the world becomes more renewable-energy friendly.

## Recommendations

The environmental impact of aviation is evident but not particularly dramatic when compared to other forms for industrial products. This is not to say that lowering the impact of harmful gases should not be a priority for aviation designers. However, the immediate task should be to make sure that current conventional technologies are made as green as possible rather than depending on electrification to clean up the aviation industry. Given the sheer number of people that airplanes currently ferry across continents and within countries, it will be a long time before EA will make even a small dent in this transportation picture. In the mean time, incremental changes like cleaner engines and ME aircraft can help. The aviation sector has got a bad rap when it comes to pollution because, on an individual basis, it is rather dramatically ahead of the pack. The solution to this, at least for the foreseeable future, is an improvement in conventional technologies.

Having said that, we certainly believe in the philosophy that every little bit helps. For example, new forms of transportation such as air-taxi services powered by AE aircraft could help reduce pollution in urban environments in small ways. Given all the activity in the sector, it does seem like these machines are going to be a critically important part of our future, but if we are depending on EA to dramatically change aviation's impact on the environment, then we are looking at a fool's paradise. Being an unashamed technologist, however, it is clear to me that the excitement that electrification has injected into our staid old field is to be cherished and encouraged, and its ability to inject some new

**FIGURE 5.** **The comparison of $CO_2$ output of engines compared to the average grid.**

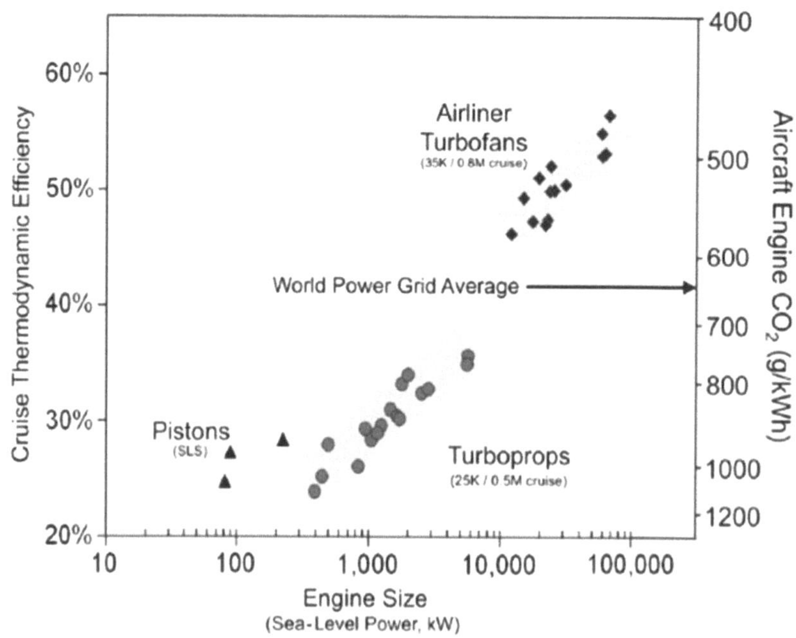


Courtesy of Epstein and O'Flarity [8]. Reprinted by permission of the American Institute of Aeronautics and Astronautics

blood into the sector more than makes up for its overall lack of big gains. Suffice it to say that its psychological benefits are far more important than its technological benefits.

Because electrification is so intimately related to the electric grid infrastructure, another recommendation is to accelerate the move toward making the grid green. That is, however, not a topic that can be tackled in this report, and many billions of dollars are being spent around the world to make the share of renewables a more sizeable portion of the power generation capacity. But for this to work, large countries like the United States, China, and India must adopt these goals on a war footing and do their best to convert their power infrastructure by moving away from coal and other dirty energy sources to a bigger fraction of renewables. Unless this is done, no amount of aviation electrification will help except in a very small way.

# Battery Systems

When asked at a recent aerospace conference, Mark Moore, Uber's Engineering Director of Vehicle Systems, clearly pointed to battery systems as being the weak link in the chain. Of the major subsystems that constitute the elements of the electric propulsion-viz. generation, storage, distribution, and usage-it is storage that takes up the most volume and weight. The most viable electricity storage technology currently available is that made up of lithium compounds. In fact, the invention of lithium batteries is so important to the life and economy of the world today, that in 2019, the Nobel Prize committee bestowed the chemistry award to the three leading pioneers of this technology, John Goodenough, from the United States, M. Stanley Whittingham, from the United Kingdom, and Akira Yoshino, from Japan, "for the development of lithium-ion batteries."

Lithium-ion secondary (rechargeable) batteries are being used extensively in ground electric vehicles and hence are making steady progress in terms of their technology, and cost-effectiveness. But for aviation applications, this progress has not gone far enough and certainly not fast enough. Let us make some comparisons.

It is often stated that the maximum specific energy density of lithium-ion batteries is about 220 Wh/kg as compared to that of jet fuel, which is about 12,000; and this roughly 55-to-1 ratio is presented as proof that AE propulsion is infeasible. While on the face of it, this does look damning, examining this claim further shows us that some adjustments need to be made to this comparison.

Comparing the all-inclusive weight of a battery system with just the weight of the fuel, without regards to the weight of the engine, makes no sense. For the right apples-to-apples comparison the totality of the systems needs to be considered. While the hydrocarbon wins out by orders of magnitude even then, the difference is not as stark. In fact, for a single-aisle class of aircraft (e.g., the A320) the equivalent specific energy of the fuel and the turbine is about 4,300 Wh/kg. Doing the same calculation for the battery system, i.e., including the motors and the power electronics to the total weight one gets an equivalent specific energy of about 180 Wh/kg. This ratio between the Jet A-fueled engine and the lithium-ion battery-powered motor is still about 25-to-1, which puts these AE systems at a big disadvantage when powering an aircraft of this size. This is a major reason why battery-powered propulsion is only viable for small aircraft and that too for relatively short flights.

Another option for a rechargeable energy is supercapacitors, which, while not delivering as much specific energy, have the potential to deliver more power. It is possible to envision a scenario where they help boost power during high-demand situations. Of course, these devices add weight, so their use has to be factored in as part of the overall design [21].

While commercially available, battery chemistries have rather low specific energy densities, there are some chemistries, still in the experimental stages, that promise much better performance. Take lithium-air or lithium-sulfur batteries, for example. The lithium-sulfur battery has been around since the 1960s and is more than twice as dense as the lithium-ion batteries available today. But, for various reasons, having to do with inadequate cycle life and the need to make the sulfur terminal heavy, the Li-S battery has not made it past the advanced research and development stage; and definitely is not yet ready for commercial prime time. The lithium-air battery is even less mature. But the hope is that in the near future these can be improved to make them viable for aviation applications [21].

According to J. Ball [4] one of the issues with lithium-ion batteries is that graphite anodes, the standard in most batteries today, are a limiting factor in improving the energy density of the cells. Silicon is a better material but has many downsides, one of them being, it swells up when impregnated with ions. Amprius, a Silicon Valley startup, has overcome some of these disadvantages by developing a battery cell with nanotube silicon anode. This allows the lithium-ion cell to hold twice as much energy as the standard cell. However, there are limitations to this technology, one of which is that the number of charge-discharge cycles is severely capped. But their superior specific energy allowed these cells to be used on the Qinetiq/Airbus Zephyr, an unmanned solar aircraft that stayed aloft for 25 days in 2018, a record for endurance. Qinetiq was acquired by Airbus in 2013 [4].

It is possible that, with more research funding, some new system may break through the commercialization logjam and save the day, but that is not something we can just keep hoping will happen; a lot more work and investment would have to be injected into this field to affect any real change.

Current estimates are that GA aircraft carrying less than four passengers traveling for about 60 minutes is the limit for current AE technology. Some companies are claiming longer flights with more passengers, but most experts do not give much credence to these claims. Note that this range will be further curtailed because, to be certified for commercial passenger operations, aircraft have to carry reserve fuel to accommodate emergency situations where the destination airport cannot be used, and an alternative needs to be reached.

There is one strategy that might help with this range problem, and it was considered by at least one company-Eviation-at one time. And that is to use primary (non-rechargeable) batteries to provide the necessary reserve energy. The idea is that the primary batteries will never be used in normal operations and, because they are generally much more energy efficient, they will not occupy nearly as much space and weight as secondary batteries. For example, an aluminum-air battery can provide nearly 1,000 Wh/kg of specific energy but the aluminum plates need to be replaced once the battery is discharged.

## Safety Issues

In 2013, the aviation world was rocked by the revelation that new lithium-ion batteries on the Boeing B787 Dreamliner were susceptible to "thermal runaway," which was another way of characterizing a battery fire. Several root causes were identified, ranging for inadequate testing to manufacturing defects to lack of regulatory guidance. While the issue with this battery system was solved satisfactorily, though not optimally, the whole incident brought up the need for more regulation and industry guidance for lithium batteries. Especially because battery systems meant for propulsion would be far more powerful and have larger charge capacities than those being used currently for restarting engines, or for emergency backup on some aircraft fleets. It is interesting to note that SAE International has identified this to be a risk factor for the industry and has reconstituted its energy storage technical committee to focus solely on lithium batteries for propulsion in addition to conventional functions. This committee has been designated AE7D and is titled "Aircraft Energy Storage and Charging Committee."

There are a number of issues related to lithium batteries that are not present in older lead-acid or nickel-cadmium ones; lithium batteries tend to have a bigger propensity to catch fire if their charge and discharge rates are not managed correctly. This means that the battery management system (BMS) is crucial to their operation. Lithium batteries are prone to thermal runaway due to a variety of reasons. But, according to Norm Pereira, a senior FAA official who has been working on developing standards for battery systems for a while, some of the known issues are

- manufacturing defects
- impurities
- foreign object damage
- damage during transport
- growth of dendrites and lithium plating accelerated by certain operating conditions

Much of the battery characteristics, and risk profile, depend on the chemistry, additives, and design characteristics of the lithium batteries. For example, lithium - cobalt oxide battery cells used in most electric vehicles, and the Boeing B787, have a higher specific energy when compared to lithium - iron phosphate cells, which are increasingly being selected for more modern aircraft installations. The downside of the $LiCoO_2$ batteries is that they are susceptible to catching fire if they are pierced. According to some reports this was the cause of some Tesla car fires. The $LiFePO_4$ cells are less energy dense, but generally are safer for critical operations. Another design choice, according to Pereira, is the use of modularity. If each cell is relatively small in size, and the entire battery system is made up of packs of these cells in series (for voltage) and parallel (for current), then the risk of starting a thermal runaway reaction is much less than for a battery system made up of larger cells. Overcharging could be a cause for the cells catching fires, but fires may also be started by physical damage to these cells. Some of the fires on electric vehicles were caused by foreign objects penetrating cells. It is for this reason that as part of the safety testing of these cells, manufacturers are required to demonstrate that the cells can survive penetration with external objects like nails.

Strict criteria for demonstrating battery safety have been developed and published by the RTCA in the form of a standard called DO311, developed by a team of industry experts. This team includes personnel from industry as well as from the regulatory agency. Norm Pereira is part of this team and speaks about these topics with authority. As he correctly points out,

> *...these lithium batteries contain very large amount of energy in a small profile. Therefore, the safe design usage and mitigation of these very beneficial batteries in an aircraft environment is critical to [their] success...*

It should be emphasized again what a key part of a lithium battery design the BMS is. Unlike older chemistries such as lead-acid or Ni-Cad, lithium tends to be riskier. Balanced charging and discharging of cells, safety monitoring, and diagnostics and prognostics are all part of the BMS function. Because this is so dramatically different from older systems, and because high-power lithium-ion battery systems are so rare even today, not much experience exists within the aerospace community in the design and application of BMS systems. This situation will, of course, change as the industry matures, but this is another area where more research and development would help. As mentioned above, the SAE technical committee that has been set up to develop documents on lithium batteries (AE7D) has taken up the challenge to produce an information report on the BMS, as one of its first publications.

## Charging Standards

All electric systems need to be charged before flight. Some form of charging system needs to be provided for the transfer of power from the grid to the vehicle, and it is safe to assume that this will be done via some physical interface just like with electric vehicles. A lot of the activity surrounding aircraft electrification is experimental in nature. For many of the prototype aircraft that are being designed or are being built, the primary motivation is to prove out basic flight concepts, and items such as standards for power charging are not high on the priority list. Because there are so many kinds of propulsion systems out there, even basic parameters such as bus voltages and power levels have not been standardized. All is in flux because different aircraft configurations have different requirements for power and reliability.

Because recharging the batteries by plugging the vehicle into a charging system can take time, it has been suggested that a better process would be to swap out discharged batteries for recharged ones. But this has its own issues, as we will examine.

Let us first examine the question of charging. In discussions with Joshua Portlock, an expert in this field, it becomes very clear that the EA industry should study the history of the electric vehicle industry to make sure that some of the pitfalls are avoided.

In the electric vehicle world, different OEMs, at different times, came up with different charging standards for their vehicles. Generally, these varied by region, with local OEMs coming up with compatible standards. For example, the CHAdeMO standard was developed by Japanese car companies for high-speed, direct current charging stations. Tesla came up with its own proprietary design; the Europeans, led by Volkswagen, developed a system called the Common Charging Standard, and, more recently, the Chinese have come up with the GB/T standard. To harmonize these competing standards, SAE has published J1772 whose latest version was released in 2017 [24]. The International Electrotechnical Commission (IEC) in Europe is doing the same, and has developed IEC61851 based on the CHAdeMO standard [13]. Because in some places there are several competing standards, the landscape is still undefined with some cars having to carry adapters to be able to connect to some charging stations.

This use of adapters is reminiscent of the case with fire hydrant connectors. An interesting fact in history is that one of the reasons for the establishment of the National Institute of Standards and Technology (NIST) was to come up with a standard fire hydrant and a hose coupling so that any fire truck could be requisitioned from anywhere to fight any fire. In the early twentieth century this was by no means the case. In the Great Baltimore Fire of 1904, fire trucks from many neighboring counties arrived in Baltimore to help fight the fire but were rendered ineffective because their hoses would not fit the Baltimore hydrants. Even today, more than a century later, some trucks have to carry adapters to be able to use hydrants in some parts of the country, where it has proven too expensive to make the changes [17].

But, coming back to the subject at hand, there are at least two basic considerations that the design of the charging interface has to take into account: the electrical parameters (voltage, current, and power), and the physical form. Additional considerations are digital communication interfaces and cooling means, though none of the existing systems use any active cooling. Today this limits the amount of power that can be safely transferred to about 400 kW. For the fast charging of relatively large batteries needed for air transportation, this is obviously not adequate. These are all considerations for the design of the standard charging system of the future. Figure 6 shows some of the different prevailing standards, courtesy of Joshua Portlock. These show the voltage and current levels and the power levels they can handle. Also shown is the rough number of installations and the communication bus protocol used. CAN stands for Control Area Network which is the most popular communication protocol in the automotive industry, and PLC stands for Power Line Communications, where the digital information is conveyed over the power conductors themselves.

The history of this can be traced back to different companies using different voltage and power levels for their vehicles. While we are starting to see some standardization with SAE and IEC publishing standards on this, whether this will completely harmonize the industry is hard to predict. There have been some recent positive developments in the industry, with CHAdeMO and GB/T groups coming together to develop a next generation of high-power chargers rated to about 900kW.

**FIGURE 6.** Direct current charging Interfaces for Electrical Vehicles.

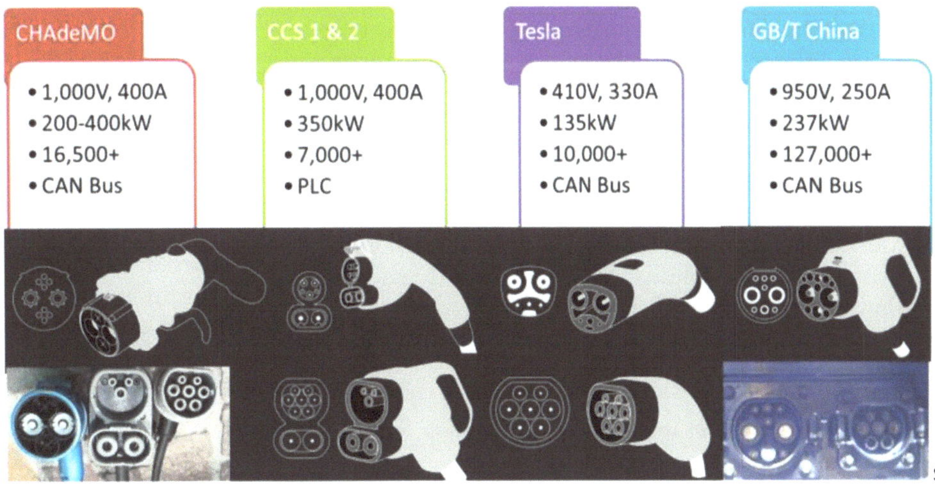


Reproduced with permission from Joshua Portlock [18]

For aircraft, the energy levels will be much higher, and it is not clear that existing electric vehicle charging systems will suffice. Also, to make sure that existing systems are not misused, aviation authorities may mandate entirely different systems for aircraft. This is still being studied in the industry.

In 2018, an SAE committee was set up just to study all aspects of lithium battery systems, with special emphasis on batteries for propulsion. Before this, the work for developing standards was the responsibility of a larger group called AE-7B which dealt with electric storage along with other disciplines such as electric power generation, conditioning, distribution, and so on. With the advent of electric propulsion, it was felt that the area of battery systems, and especially lithium-ion battery systems, deserved a committee of its own. This led to the birth of AE7D, Aircraft Energy Storage and Charging Committee. One of the first documents being developed by this committee is a standard for charging connectors. It is hoped that by getting ahead of the game, the aviation sector will not go down the route of the ground-vehicle sector where we still have a number of competing standards.

An alternate method of charging batteries is not to do so on the vehicle. Instead, just swap charged ones for discharged ones and then recharge the depleted batteries afterwards in a dedicated facility. There are pros and cons to this idea. The two strongest pros are that "refueling" the aircraft would be a cinch because it would take much less time to swap out a battery when compared to recharging it. The second pro is that the discharged battery could be charged in a controlled environment using the highest allowable power level under extremely safe conditions. One con is the logistics and economics of this system. An extensive logistics system needs to be put in place, to ensure that an adequate supply of fresh batteries is present at every location an aircraft is going to land. This will be complicated and expensive. Also, having a dedicated facility that is charging several batteries at once would require an enormous amount of power to be supplied. Another downside to this would be an additional requirement on the aircraft designer to ensure that the batteries are readily accessible to be swapped out. This would include physical access, design of adequately positioned and spring-loaded contactors, foolproof means of securing the batteries, and so on. This is not going to be easy, because of the sheer number of batteries that will be needed to deliver the energy that will carry a meaningful payload over a reasonable distance. Just like in electric vehicles today, where battery packs take up a large volume within the frame, the integration of the batteries and the airframe would obviate an easy solution with batteries that can be swapped. In addition, liquid cooling systems that will probably be needed would have to be disconnected and reconnected. Take one of the leading contenders in the bigger class of aircraft-Eviation from Israel. It has been reported that more than half the weight of the aircraft is in the batteries; much of it built into the airframe [2]. As the founder of Eviation says:

> *This is a truly form follows function situation ... it was built from the ground up to be electric ... it's a huge battery with some plane painted on it.*

There is no way the Alice, which is what the Eviation AE aircraft is called, will have swappable batteries. So, for practical reasons the only aircraft where this might make sense are very small motor gliders or training aircraft, like the Pipistrel Alpha.

## Recommendations

The biggest barrier to an economically viable AE aircraft today is the energy density of the battery system. Given how many instruments of daily life are dependent on this technology, recommending that more research be conducted in this area is stating the obvious. From laptop computers to electric vehicles to drones to EA, all are waiting for denser batteries. There are some new chemistries that are being experimented with, but nothing radically new and extraordinary seems to be on the horizon. Among other things, it would behoove research-funding agencies to emphasize this as one of the critical technologies of the future and redouble their efforts to direct funding toward finding the next generation of the rechargeable battery system. In addition to novel chemistries, accessory systems such as the BMS, which are, nevertheless, integral to these batteries, must be researched and further optimized. They can have a tremendous impact on the operation as well as the safety of these batteries.

The same, to an extent, can be said of the other elements of the electrical transmission: the power electronics, the distribution system, and the electromechanical machinery such as the generators and the motors. All of them can benefit from a weight-loss program, because in the aviation game every ounce counts. In general, the progress on reducing the size of power electronics (PE) components has been quite satisfactory, with numbers around 13 kW/kg achievable today and numbers around 25 kW/kg in sight of researchers. The picture with rotating machinery is not as encouraging, but that is to be expected because of the nature of any electromechanical component with moving parts. Today the specific power numbers are around 5 kW/kg but with some manufactures touting figures twice that. The goal is to be around 15 kW/kg. Distribution systems depend on many factors but mainly the driving factor is the amount of current that needs to be transmitted. For the same transmitted power, the current can be limited to an extent by raising the voltage levels, and there is a lot of research in figuring out just what the right combinations should be. NASA has set up an electric test site at their Plum Brook facility near Sandusky, Ohio, called the NEAT. It tests full-sized systems to validate different electric architectures for EA. It is capable of handling 24 MW of power, which is an extraordinary level for a laboratory. There are other government test labs that can help with this task as well.

The only recommendation for battery-charging systems is for the EA industry to get together and develop some standards. The lessons from the electric vehicle industry should be clear enough. Having different systems can be costly for the industry as a whole. All major players should have a vested interest in getting together now and come to consensus on systems that can be used across the world. It is encouraging that the aerospace standards organization within SAE

International has formed (AE-7D, in 2019) that is looking at various aspects of lithium-ion rechargeable batteries. Chief among these is the question of charging systems. While it is too early to say which way the committee is leaning, it is keenly looking at the electric vehicle systems existing today to figure out what to do next.

# Thermal Management

Large currents and electrical resistances do not play well together. That is why electronic devices heat up and generally need some kind of active cooling. Even low-power devices, such as a laptop, have forced air cooling to deal with the heat generated by the power supply and the internal electronics. When dealing with PE devices integral to electric vehicles or electric aircraft, the thermal issues are orders of magnitude larger. Hence, the need to manage waste heat in these systems, consisting of batteries, PE, generators, and motors, is a critical part of the design challenge. It is even more critical in EA than in aircraft propelled by hydrocarbon-fueled engines. In case of electric vehicles, liquid cooling is employed very much like in conventional vehicles, using an liquid-air heat exchanger (radiator) to reject the heat. This luxury is not available to the EA designer. The entire issue of thermal management is the subject of a lot of research and development, especially with those working with larger aircraft. It is still not clear whethere this problem has been effectively solved. The content in this section was especially influenced by inputs from Steve O'Flarity from Pratt & Whitney and Rodger Dyson from NASA GRC.

When it comes to heat rejection, one advantage conventional engines (i.e., hydrocarbon-burning reciprocating engines or gas turbines) have over other devices is that they are open-cycle devices. The working fluid is heated up and after it does its job (of moving the piston or the turbine blades) is rejected in the exhaust along with its remaining heat. This makes these engines inefficient and polluting, even if the exhaust gasses are treated properly. But on the flip side, it does obviate the need to find complicated cooling mechanisms to deal with excess heat. In heat engines that are not being used for transportation, the loss of useful heat energy can be very expensive. Typically, in power plants where the main power source is a GT, there is a bottoming cycle consisting of a steam turbine that is run off the exhaust stream of the GT. A steam generator uses the exhaust gas to produce steam that is used in the steam turbine to augment the power output of the GT. To ensure high overall thermal efficiencies, large power plants are generally configured as "combined cycle" plants. This allows the combined thermal efficiency of the entire system to be more than 60%, which is quite impressive, given that the efficiency of the GT itself is close to 40% or so. Such a bottoming cycle is not very practical for aircraft application, resulting in losses that lead to increased environmental impact. But the aircraft engine takes advantage of this very feature to manage its thermal load. There are a few more parasitic thermal loads that the conventional aircraft engine needs to contend with. Chief among them is the heat generated by the bearings. A closed-circuit oil system is used to cool the rotating components, and the heat in the oil system is generally rejected to the air via heat exchangers. Another source of heat is the electronics, generally housed in the electronics bay in the aircraft, as well as on the engine. The e-bay electronics are generally cooled with forced air, while air and fuel are used in the engine. The latter has the advantage of heating up the fuel before it is burned, which increases the thermodynamic efficiency of the overall system by some amount. But all told, the heat load of an engine for a narrow-body aircraft does not amount to much more than 100 kW for machine that can generate more than 35 MW of power [8].

This is not the case with the EA. The heat loads associated with motors and PE devices can be 20 to 40 times as much as in the case of a conventional engine [8]. All this heat needs to be rejected, either to the air directly or through an intermediary liquid cooling cycle. It is not too bad for motor gliders and smaller GA aircraft, but for larger aircraft it becomes a fairly heavy burden. Because this is a well known but as-yet-unsettled issue, it is an active area of research. While this applies to all sizes of aircraft, the burden is especially high for larger systems, hence teams that are looking at larger aircraft have this as one of the critical priorities.

When the aircraft is in motion, ram air heat exchangers can be employed to reject heat to the airflow but this solution adds weight and aerodynamic drag. In fact, NASA has done extensive studies on the different kinds of cooling systems that may be needed to solve this problem, and none seems to be adequate [6]. A number of systems have been proposed, as outlined by Dyson [6, 7]. All of them have their limitations, and even more exotic solutions are being proposed to tackle this problem. One of the more exotic of these is a cooling system that depends on acoustic energy to accomplish this heat transfer. Details are available in some of the papers written by Dyson and his colleagues [6, 7].

Another means to tackle resistive heating is to not generate it in the first place. Superconducting systems have close-to-zero resistance, and hence generate little or no heat. The idea is to cool the working materials to below their critical temperatures so that heat is not generated in the first place. However, this brings with it a plethora of other problems.

Another issue with running electronic devices at elevated temperatures is that this is not very beneficial to the life of the devices. Even if the temperatures are managed during flight, there are long-term effects associated with running equipment at high temperatures, such as component degradation. Work that we are doing with power-switching devices show that remaining useful life is critically affected by operating temperatures. Integrated gate bipolar transistors are the heart of switching devices called invertors that are needed to control motors in EA vehicles. Much work has been done in recent times to assess the health of these devices and to develop prognostics algorithms for them. It is clear that the temperature at which an integrated gate bipolar-transistor device works is a critical parameter in determining its life. This is another reason for ensuring that cooling is a critical part of the design.

## Is Cryogenics the Answer?

One of the suggested routes for increasing electrical efficiency and reducing the size of electrical components is to introduce superconductive components into the transmission. This can be done by ensuring that all critical components operate at reduced temperatures where the current-carrying capacities of conductors is increased exponentially, thereby nearly eliminating the losses due to electrical resistance. Ideally, these conductors would operate at supercooled states of near 0°K but that is practically impossible on a commercial scale. As a consequence, what is being talked about these days is high-temperature superconductive materials that show remarkably low resistances at relative higher temperatures of around 50°K to even 200°K. Still there is a need to be able to generate these low temperatures using some means of refrigeration, or to be able to carry around tanks of cryogenic liquid that is used up during the flight. There are obvious issues with both, which we will get to in a minute. The biggest concern is related to safety. If there is any accident that results in a breach in the system, whereby the superconductors heat up-even for an instant-the results can be catastrophic. This means that the certification issues associated with such a system have to be dealt with in the most rigorous and comprehensive manner. Along with other certification issues related to EA that have been alluded to elsewhere, this becomes a much bigger design challenge. That is one reason such a system is only currently being studied as a research topic in government labs such as NASA and major OEMs like Airbus.

Regardless of this, what is the feasibility of cryogenic systems? The discussion below has been influenced by discussions with Rodger Dyson at NASA, who is a leading researcher of such systems. He and his colleagues have set up a major test and validation facility called NEAT to study realistic EA architectures at their Plumb Brook facility in Sandusky, OH. NASA's Glenn Research Center has been a leader in the study of EA technologies and have extensive test facilities at their main campus in Cleveland, OH. In fact, as described in a recent paper by Hartwig, et al. [11], NASA has set up extensive testing rigs, which allows them to test electrical properties of systems at cryogenic temperatures. The paper's conclusion, quoted here in its entirety pretty much sums it up:

> *This paper outlines the test capabilities, test parameter ranges, and details of each of the testbed subsystems, including test articles, core rig, power, fluids, instrumentation and DAQ, and control subsystems. A new LH2 based testbed has been designed to measure AC losses of superconductors between the range of 18-28 K using LH2, but easily extensible to 95 K using GHe or LN2 as a working coolant. While the target application is 20 K $MgB_2$ stator coils, the rig can accommodate low-TC as well as high-TC superconductors. The rig can handle both AC as well as DC testing, perform testing with no-injection (induced current only) as well as injection current testing, and accommodate a plethora of different sized and shaped wire, tape, cable, and coil test articles. The system is designed for the following test parameters: injected current (0-400 A), frequency (0-400 Hz), magnetic field strength up to 0.6 T, phase angle between induced voltage and injected current (-180-180°), and AC power loss (5-30W). The new spin rig builds upon the existing spin rig capabilities in several ways as outlined previously.*

Magnesium di-bromide ($MgB_2$) is the primary material constituting their first tests. $LH_2$ and $LN_2$ refer to liquid hydrogen and liquid nitrogen, so we're talking very-low-temperature systems. The question for the practical system designer is: Given the extensive infrastructure needed to support superconductivity (even high-temperature superconductive materials) in the air, is it really worth it? We know that as we get bigger and bigger aircraft (B737-class and higher, which is what the bulk of the flying public uses today), do the thermal balances make sense for cryogenic systems?

A superconducting power electronic system, for the same wattage, will be much lighter than one operating at ambient temperatures. And it will generate less heat. So, we gain a huge weight advantage, by making the PE components smaller, and because the systems engineer has to install less heat rejection equipment. But, on the flip side, we need to either carry tanks of cryogenic fluids or have refrigeration systems to continuously cool the working fluid (possibly liquid hydrogen or liquid nitrogen). Also, because of potentially catastrophic conditions arising from these systems suddenly being exposed to ambient temperatures, we need to build safety systems that contain any thermal runaway and guarantee that passengers are not harmed. In order to reduce the probability of catastrophic failures at the system level, redundancy will have to be built into the system to at least ensure some kind of "limp-home" operation. This in turn will lead to a big increase in weight and volume. It is not clear that all the necessary system-level studies have been done to ensure the feasibility of such a system.

Researchers at NASA have been studying the economics of having a HE fuel-cell aircraft where $LH_2$ is the working fluid. It is first used to keep the superconducting electrical components cool below their critical temperatures and then is used in the fuel cell to generate the needed electrical energy. Whether this is feasible from a practical point of view is still to be determined. If such a system is finally designed for commercial use, the fact that potentially hazardous hydrogen is being carried around must figure in the design as well, and its risks mitigated. They have kicked off a three-year program with University of Illinois leading the effort to develop some concepts for this hybrid system (Figure 7).

## Recommendations

The challenge has been stated very succinctly by O'Flarity:

> *Any conventional method of transferring heat outside the aircraft will negatively impact drag and efficiency. New ideas must be pursued, either to utilize it in some beneficial way or reject it with minimal loss.*

**FIGURE 7.** **Concept for a $LH_2$-based fuel cell-powered electric aircraft (University of Illinois).**

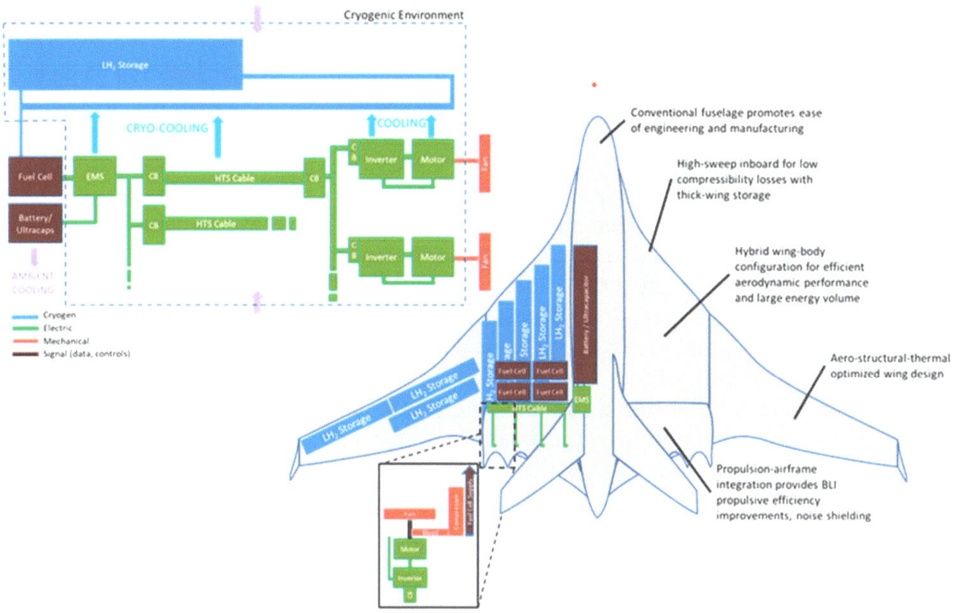


Reproduced with permission from CEC-ICMC Conference Hartford, CT (July 21-25, 2019) [30]

Given the dramatic increase in thermal loads that EA will experience, there is really nothing to recommend other than to say that this must be a critical part of the design of the system. The work that is being done by agencies like NASA is extremely important in developing new cooling and heat-exchange technologies. If any of these become technically viable, they could dramatically help with the problem of designing a highly efficient, but lightweight package to take care of the thermal loads inherent in electric propulsion.

Cryogenics and superconducting electronics are great ideas and the biggest future programs seem to rely on them, at least if the current designs are to be believed. But the overall efficacy of the system has not yet been proven. NASA's Cleveland and NEAT facilities have the infrastructure to develop such systems and test them out. It is likely that something will come from all this research in the future, either proving that this is a good idea or showing why it will not work. The idea of coupling a $LH_2$ system with a fuel cell is really intriguing, and one hopes that the work by the University of Illinois and its partners shows how this can be made into a reality, if at all.

## System-Level Issues

A normal car engine, not built for the racetrack or the high-end luxury market, can develop around 150 kW of peak power without burning itself out. This translates to about 1.7 kW/kg of specific power, making generous assumptions on the weight of the engine. By contrast, the specific power (at takeoff) for a GT in the CFM56 class would be around 4.1 kW/kg. There is a reason why GTs have been around for nearly 70 years of active service; ounce for ounce, they are one of the most effective air-breathing propulsive machines built. Yes, a rocket, say for the space shuttle, can generate around 150 kW/kg, but that is not an air-breathing engine. While they have serious downsides, like a much higher environmental impact, it is difficult to find a comparable means to propel an aircraft through the sky. Given this reality, what is the world to make of this electrification revolution? The bottom line is that for all the reasons we have outlined above, it is not going to be an easy sell. In an expert panel at an AIAA forum in August 2019, Michael Winter from Pratt & Whitney made some very insightful comments about energy and the environment and what the future of electrification means. The key point he made was that it is not a contest between the conventional cycle or the electric propulsion cycle. The goal is to optimize the entire ecosystem to make sure and figure out where each technology fits. Reporting on work that his company has done, Winter said that there is a lot of runway left in the Brayton cycle-based jet engine. An interesting graph he shared was one that showed how much room there is in increasing the efficiency using just the standard cycle, if systems could be built that could withstand the temperatures needed. Figure 8 shows this graph from Koff [16] and even if one looks at current systems that are about a little less than halfway up the curve here, there is still a lot of growth potential for GT technology. This is only talking about core efficiency and does not consider the entire efficiency of the engine, which includes the propulsive efficiency of the fan, but the point is well made.

Following up on his point about a diverse future for aviation, Winter said that air cargo accounts for close to 35% by value of all cargo delivered around the world, even though by volume it only accounted for about 0.5%. With the growth of e-commerce, this sector is poised to explode. We have all heard about the experiments that Amazon and others are doing in bridging the last mile using unmanned eVTOL aircraft (i.e., drones). What is less well known is all the great humanitarian work that drones are doing, especially in areas of the world where transportation infrastructure is inadequate; services such as delivering vital and perishable medicines.

Given the potential illustrated in Figure 8, it is easy to envision a future where emissions for the GT is 15% lower than what it is today, just because of the fact that it will burn that much less fuel. This was echoed by the other members of the panel: Eric Ducharme from GE and Mark Wilson from Rolls Royce Liberty Works. Pratt & Whitney has, of course, pioneered the concept of the geared turbofan, wherein the fan and the low-spool are decoupled by the means of a gearbox, allowing the fan to run a third slower than the much faster low-pressure section of the turbine. The increase in propulsive efficiency has resulted in the geared turbofan achieving greater than 15% lowered fuel-burn numbers compared to that of the older class of GT engine for the single-aisle narrow-body aircraft. GE (and its partner in the CFM joint venture, Snecma) has achieved similar results by cranking up the pressure ratio in the compressor, and thereby the highest temperatures achieved in the hot section. They are using lighter and more exotic materials such as ceramic matrix composites in key areas on the turbine, and the fan is fully composite. The thermodynamic efficiency of the Brayton cycle is critically limited by the highest turbine inlet temperatures in the GT. These design innovations have resulted in higher efficiencies in the LEAP class of engines compared to the older CFM56 class. The hope for the future is that by marrying the two technologies (higher propulsive efficiency and higher thermal efficiency) even better fuel-burn numbers can be achieved. Rolls Royce has been working on interesting technologies of its own, including geared fans that can vary the pitch, which is called UltraFan, and which would allow the thrust reverser to be eliminated.

The fact that the growth in the evolution of the GT is not complete does not mean that engine OEMs do not have to consider the threat from the whole electrification movement. If in the future this becomes a roaring success, at least for the single-aisle aircraft around the 50 to 100-ton grow takeoff weight class, the GT will become a minor piece of the propulsion equation. It will be a relatively small unit running at a constant speed powering a generator with the electrical energy stored in a battery. This will then be used to drive a distributed set of fans providing propulsive power. It is not clear that for much larger (twin-aisle) class of aircraft the HES would be the right solution, but at least for the medium single-aisle class it is entirely possible that the HES would be a viable highly efficient solution. Today, such a GT is used as the auxiliary power unit in the aircraft, providing cabin air conditioning, driving the generator, and providing air for engine start. In a hybrid future, will the jet engine be relegated to serving this function? In this case will the turbine OEM just become another supplier to the airframe OEM without any special status in the aviation hierarchy? It is possible that the special Part 33 designation for engine certification will disappear and the GT will become another chapter in the Part 25 requirements. All this is speculative, and the scenario is probably not likely to happen, but for GT manufacturers, it is certainly worth looking at. Richard Ambroise, from Airbus, has been considering these questions for a while since he is an expert in the area

**FIGURE 8.** **Improving core efficiency in the Brayton cycle.**

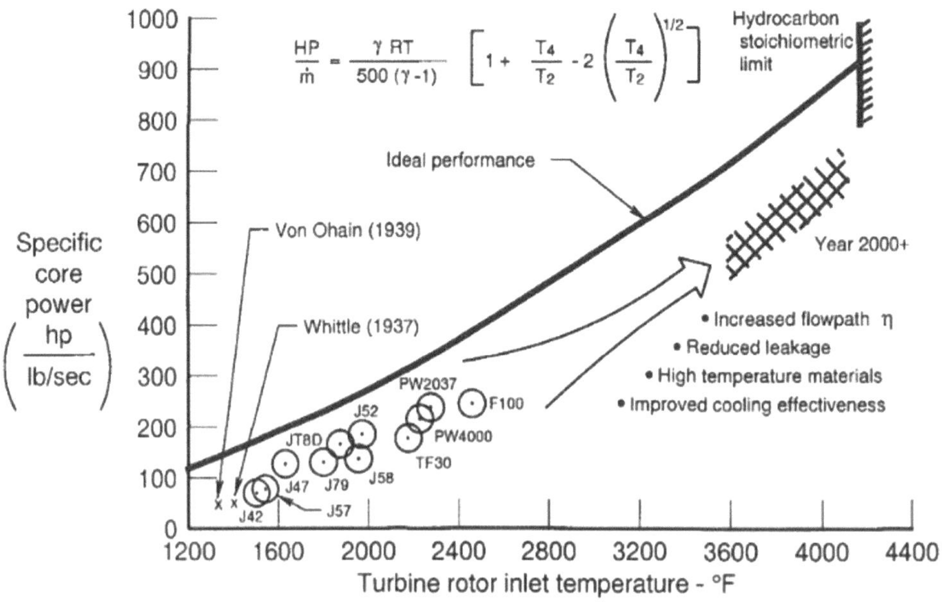


Reprinted by permission of the American Institute of Aeronautics and Astronautics, Inc. [16]

of engine and aircraft integration. He is currently the chair of a newly formed technical committee of SAE International on Aircraft Hybrid/Electric Propulsion (E-40).

This existential threat is one reason engine OEMs are busy developing their own EA programs. United Technologies Corporation recently announced its Project 804, which consists of retrofitting a Bombardier Q100 (derived from the Dash-8) with hybrid propulsion; an HEP to be precise. They are hoping to achieve a 30% increase in efficiency using this configuration. Rolls Royce, for its part, has been working with Airbus over many years on European programs to develop future HE aircraft. They are part of the E-Fan X program along with Airbus and Siemens to convert a British Aerospace 146 to a hybrid configuration. One of the four Honeywell engines will be replaced by a Siemens 2MW motor and the electric energy for this will be produced by a fuselage-mounted Rolls Royce (Allison) AE2100 turboshaft engine driving three (reportedly, Honeywell) generators. In concert with its ambitions in the electrification field, Rolls Royce announced in June 2019 that it was buying Siemens' eAircraft division, based in Germany and Hungary, that has been developing and testing electric propulsion, avionics, and control systems for light aircraft.

As far as this author can tell, GE has not explicitly made any public moves in developing an EA, but it is working closely with NASA to develop efficient and light PE and electromechanical components. In September 2019 they kicked off a NASA GRC-funded program to develop an efficient invertor-a key to EA propulsion-using silicon carbide technology.

## Ancillary Systems

Finally, we need to talk about some other systems-level issues that are far from being resolved. Figure 9, shared by Professor Craig Lawson, depicts all the systems on the conventional aircraft that can be subjected to electrification. This is the evolution of the ME aircraft. As can be seen, all systems are being considered for conversion. Two of them are of particular interest to us: the actuation of primary and secondary control surfaces, and the ice-protection system. Both use up a lot of energy and both have been part of the ME transformation of traditional aircraft.

Today, most ice-protection systems that heat up the leading edges of flight surfaces and engine inlets are based on hot air generated in the jet engine. On modern aircraft, such as the Boeing B787, where engine bleeds are no longer available, these are being converted to electrical heating systems. These heating systems can mess up the overall energy balance of the aircraft and provision must be made for powering them. This will be another drain on the electrical system, subtracting from the power needed for flight.

**FIGURE 9.** **Increased electrification of the aircraft.**

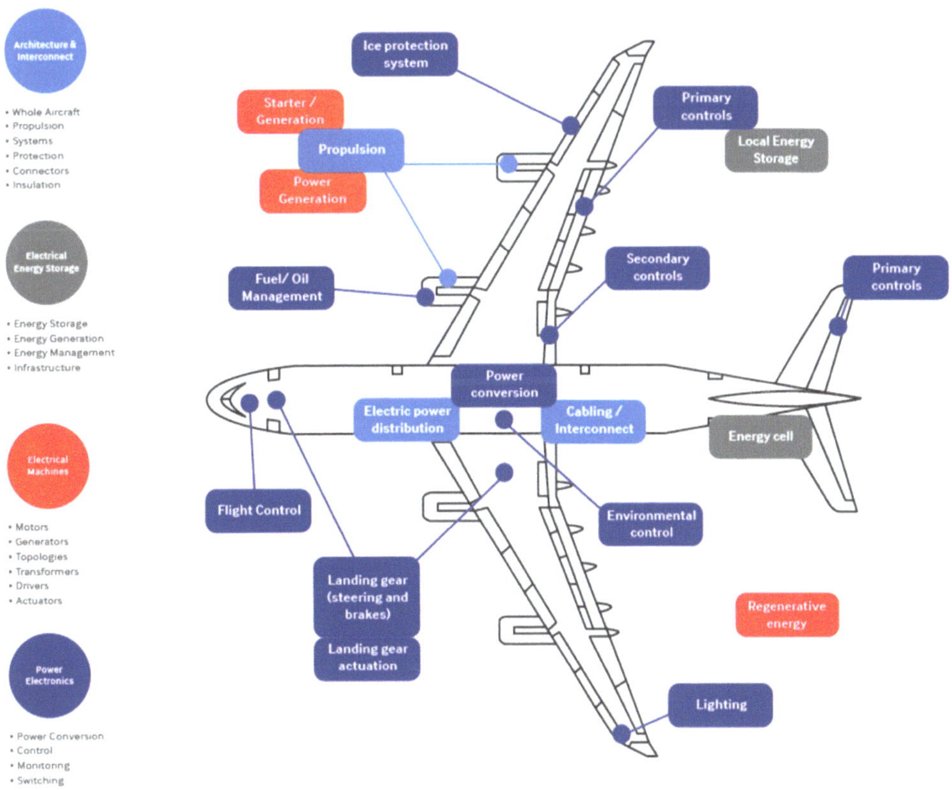


Reproduced with permission from Aerospace Technology Institute

The actuation of control surfaces is a key element of any aircraft. Today, actuators that power critical control surfaces are driven by hydraulics, partly because that is the way they have been done forever, but mostly because hydraulic systems are reliable. The mechanical ball screw that is integral to an electromechanical actuator (EMA), which would be the logical replacement for the hydraulic system, would make the system less reliable because it could jam during operation. Currently, that is one reason these actuators are not approved for primary control operations, only for secondary controls. Having said that, there are two applications of electric brakes-employing just these EMAs-that are flying today, the B787 and the Bombardier (now Airbus) C-series aircraft, Airbus A220. These get around the reliability issue by having very high redundancy. This is not possible for future EA because of the unacceptable increase in system weight. However, the entire issue of control surface actuation may go away with distributed propulsion that is being envisioned for some EA systems. With many smaller propulsors, positioned in strategic orientations, the need for control surfaces might just disappear!. The helicopter is a prime example of an aircraft which does not use any actuated flight control surfaces. The main rotor provides all the control. Some of the smaller EAs, like the Volocopter and the eHang, are essentially helicopters. In the future, novel architectures may include multiple fans and propellers that obviate the need for surfaces for maneuverability.

However, that is a future scenario. For the time being, we are stuck with EMAs. Tremendous progress has been achieved in making EMAs more reliable; and it is entirely conceivable that in the not-too-distant future we will see them serving in critical areas. It is just that for EA applications, it is absolutely critical that electrically driven actuators be made as reliable as their hydraulic counterparts. Carrying an extra hydraulic system on an EA would be ridiculously inefficient. So, one of the essential prerequisites for being able to certify EMAs to operate safety-critical control surfaces would be to ensure that they are as reliable as current actuators. The SAE technical committee HM-1 on integrated vehicle health management is working on a document on the diagnostics and prognostics of EMAs that would help support this goal of increasing their reliability.

Finally, it is not clear that anybody has given much thought to developing procedures on how first responders will deal with the crash of an EA. Even a small EA, carrying one or two persons, will have a lot of electrical energy stored in it. The crash of an EA could result in serious injuries to first responders if response procedures have not been well established, published, and personnel trained. The work being done now to develop such processes for ground-based electric vehicles will help with this task. It is hoped that experts in the aviation industry are talking to their ground vehicle counterparts to learn from their procedures and experiences.

## System Certification

This is not an easy subject, and much remains unsettled. This is not the forum to discuss it, because the issues are very complex. At least two different organizations are working on different sets of standards for EA. While SAE International is concentrating on the larger-categorized as Part 25-aircraft, ASTM has taken on the challenge of developing guidance for smaller GA class of EA seating 19 or fewer passengers (Part 23). The ASTM committee F44 on Aircraft Systems is spearheading this. Gilles Rosenburg, who is involved with both ASTM and SAE, is the cofounder of Faraday Aero near Paris, France, and he makes some very interesting points about system certification. His view is that it must be an incremental process. Also, there is scope for independent suppliers that can leverage lessons learned across different programs which a small OEM might not be able to do. Smaller startups have realized this and in some cases are hiring industry veterans to provide systems engineering and certification support. Companies like Airbus, Boeing, and Collins have an advantage in this area because they can draw on internal resources, who can step in and help their EA divisions.

## Financing

Another unsettled area is financing. There is a buzz about the EA sector that is a little unhealthy and a little disconcerting. In 2017, Lilium, a developer of an AE eVTOL aircraft, announced that it had secured $90M in venture capital funding, with money coming from sources such as the Chinese internet giant, Tencent, in addition to traditional venture capitals. Wealthy individuals like Larry Page have self-funded some adventures, while large aerospace giants like Boeing, Airbus, and Collins Aerospace are all working on their own electric programs or are acquiring companies that have such programs. For example, Aurora Flight Sciences, a U.S. company, was bought by Boeing in 2017, which helped Boeing leapfrog over many other startups in the small AE aircraft sector. This was right after Aurora had demonstrated the flight test of a scaled version of their Defense Advanced Research Projects Agency (DARPA)-funded Lightning Strike eVTOL aircraft. Even though that program was shut down by DARPA, there is no doubt that development continues within the company.

There is a lot of government support for large EA programs from the likes of NASA in the United States and DLR in Germany. Funding is being directed to private companies and universities to advance component technologies as well as entire systems. As mentioned above, these government entities are also setting up extensive labs to support the products arising from all this funding, and to support their own internal research efforts. Dubai seems to be going ahead with its air-taxi experiment with Volocopter. The idea is for the two-seater VC2 to be used in autonomous mode to ferry people to and from designated points in the city. Initially, it will be piloted but will become unmanned when its autonomy has been proven. Dubai is developing rules and regulations for certifying the craft for urban flight and UAM services. Uber Elevate has grand plans to do something similar as an extension of its Uber service. It had plans to test its UAM model in large metropolises. Not sure exactly where we are with this grander plan today, but Uber seems to be going ahead with trying out mixed-mode transportation venture by coordinating car and helicopter service in New York City, where people can order a combined trip involving cars and helicopter to JFK. This is not quite UAM, but it is a start. Uber is not the only one offering such a service around the world; but is the most well-known! They have recently announced a partnership with Hyundai to develop an eVTOL aircraft, in addition to all their other collaborations with aerospace companies.

The tremendous explosion in interest in electric propulsion has not gone unnoticed by standards-making organizations around the world. As mentioned, ASTM has been working on developing certification guidelines for smaller EA. SAE International has so far concentrated on larger aircraft but, because of its automotive legacy, is working on reciprocating engines for smaller aircraft as well. In 2017 it set up the Electric Aircraft Steering Committee, chaired by Pascal Thalin and spearheaded by SAE Staff Engineer Mark de Angelo, that looked at all aspects of electric propulsion, mapped out all the gaps in standards and guidance and is recommending actions that SAE and other organizations should take to bridge these gaps. One of the gaps that was identified, for example, was that there was no technical committee for working on electric propulsion documents. This led to the creation of E40, with Richard Ambroise, from Airbus, as the founding chair. A second recommendation was that a special committee on lithium-ion storage be set up, which resulted in the forming of the AE7D committee. Josh Portlock is leading a team within AE-7D that is developing standards for electric charging for EA. The hope is that, as these technologies mature, the Electric Aircraft Steering Group will only be needed to occasionally provide an oversight function and the need for day-to-day activities will lessen.

# Summary/Conclusions

It is clear that many aspects of electric propulsion for the aviation sector are unsettled. Only a few of the key elements have been considered here. There will be other opportunities to add to this list; the entire field is in flux as players come in and go with no clear picture about what the successful business model will win out in the end.

In addition to some of the major technical issues, we have also discussed issues such as financing and support of the enterprise and system certification. We know there are severe limitations to the technologies and the business models, and that players will struggle to make these ventures profitable. But that is not deterring people from trying their luck. It remains to be seen whether and how they will succeed.

NASA announced a 4-year Urban Air Mobility Grand Challenge (UAM-GC) in the fall of 2019. This program, starting in 2020, will try to prove, with explicit flight tests, how UAM systems will work with real-world conditions (Figure 10).

As we mentioned earlier, there are already close to 200 development efforts around the world for designing or building EA according to Roland Berger [27]. It is obviously too early to say which ones will survive, especially given the fact that the playing field is not even. A few companies

**FIGURE 10.** **An artist's rendering of the NASA UAM-GC which will kick off in 2020.**

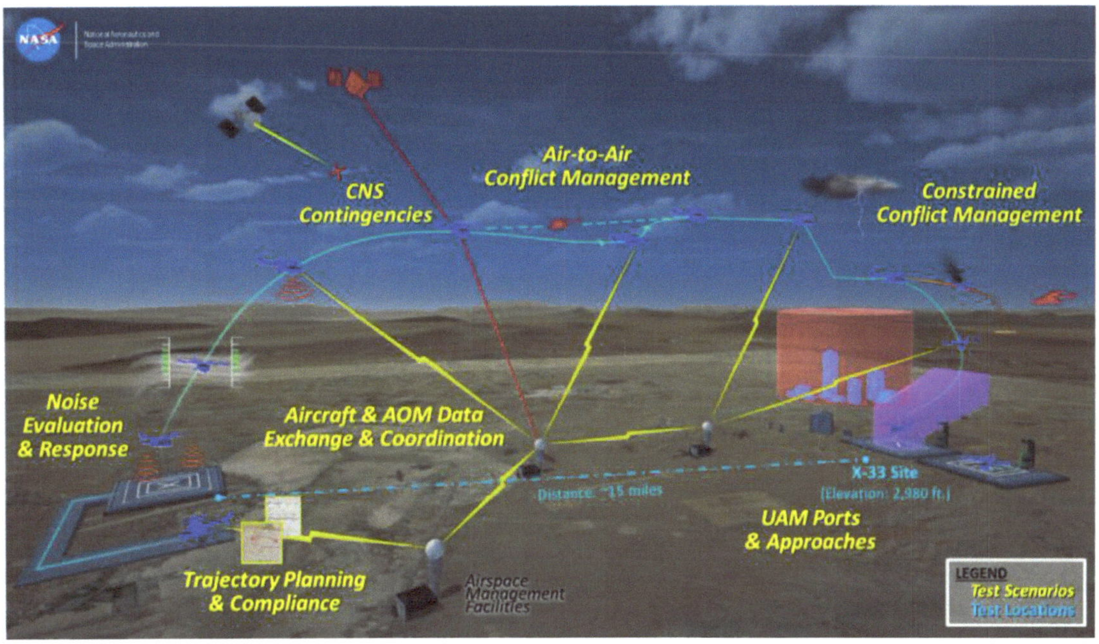


NASA

are heavily subsidized by government research grants, so their downside risk is low. These tend to be large companies who have the wherewithal to withstand temporary setbacks. Smaller firms, while far more agile, may not have this resilience.

Is there something in these technologies that these venture capitalists and funding agencies see that we cannot? It is not at all clear; given all the issues we have laid out in this report, what it is that motivates people with deep pockets to support relatively inexperienced startups and other EA entrepreneurs? Are they waiting for governments around the world to mandate such drastic emission standards that the only solution is going to be a wholesale conversion to electric technology? Or is it just the irrational exuberance of working in a truly new frontier of a cool technological sector that is the prime motivating factor driving these investors? Maybe the answer to these questions is right around the corner, or we may have to wait 25 years or more to find out. Which time horizon you feel comfortable with depends to some extent on whether you believe the more enthusiastic proponents who are relatively new entrants to this field, and therefore are not biased by conventions. Or whether you believe some of the more knowledgeable and experienced experts on the subject who, by that very nature, are more cautious.

## Next Steps for Electric Propulsion

This publication should be considered only as a first step toward clarifying some of the outstanding issues surrounding electric propulsion. The intention behind this and other SAE EDGE™ Research Reports is to start a dialogue among interested parties on important industry-wide topics that require further attention. The expectation is that these explorations of unsettled areas of technology will lead to the formation of working groups and, ultimately, committees that can address and resolve the issues they raise, producing a framework for developing a common vocabulary of definitions, best practices, protocols, and standards needed to support continued progress toward safer and more innovative automotive products.

The original impetus behind the publication of this report is that even though much progress has been made in electric propulsion, there is still a lot that needs to be done to settle issues related to this technology. What happens next within the industry, government agencies, and standards organizations such as SAE International and ASTM will be crucial in determining whether this industry will succeed in achieving its touted potential. SAE International has demonstrated its lead in this and closely related areas by setting up the Electric Aircraft Steering Group and related technical committees. These will be key determinants in how this industry evolves as they rally the industry behind them in developing standards and guidelines for each aspect of the technology.

This SAE EDGE™ Research Report identifies the following key topics for further pursuit, both through continued informal discussions among industry practitioners and through more formal working groups:

- The energy and environmental conundrum
- Battery systems
- Thermal management
- Other issues at the system level

## Recommendations

The overall recommendations of this SAE EDGE™ Research Report can be summarized as follows:

- Emissions from the aviation sector do not contribute significantly to global warming when the transportation sector is considered in its entirety. However, on a passenger-mile basis it is by far the worst culprit. Electrification cannot solve this problem for a long time because the category of aircraft involved (medium to large) do not lend themselves readily to this technology. Programs to reduce the environmental impact of current technology should be encouraged in the meantime, and we should not hope for EA to come and save the day.
- Improving the condition of existing power plants so that they become less polluting or replacing them on a war footing with green alternatives, must be priority number one for governments around the world if they are serious about combating global warming.
- More research must go into battery technology to improve their power density. Because battery technology affects so many diverse field, this is one of the major areas of research in the field. So it is entirely possible that the desired goal of quadrupling the energy densities will be achieved sooner rather than later. Any such dramatic change in the performance of the battery system would completely change the landscape for electric aircraft. Until that time, we have to live with HE aircraft or shorter flights.
- Charging standards for electrical aircraft should be globally harmonized so that we do not run into issues similar to what we did with the ground electric-vehicle systems.
- Thermal issues need to be considered as a priority in the design of EA, for the larger class of vehicles. Without adequate heat rejection technologies, this class of EA will not be viable. New technologies, like superconductivity, should be studied but within the context of newer architectures where the overall systems energy balances work out favorably. Working on these novel concepts without a systems engineering point of view will not result in successful architectures.

- Sooner, rather than later, certification guidelines need to be put in place. It is encouraging to know that organizations such as SAE International and ASTM are taking positive steps to bring this about. It is also encouraging to see that these efforts are being supported enthusiastically by the industry and the regulators.
- AE propulsion is probably not going to be economically viable for a very long time. The "killer app" of this genre is thought to be air taxi, but there are several issues with this model. Without realistic tests of these systems, their viability will never be proven, so more such proof-of-concept tests should be carried out. It is encouraging to see the experiments in Dubai, the moves that Uber is making, and also the new NASA UAM-GC program. These will only help to make electric propulsion a more credible venture.

## SAE EDGE™ Research Reports

SAE EDGE™ Research Reports, like the present report on "Unsettled Domains Concerning Electric Propulsion Technology for Commercial Aircraft," are intended to push further out into still unsettled areas of technology of interest to the mobility industry. At times, SAE launches these reports before attempting to form a joint working group, let alone a cooperative research project or a Standards Committee. At other times, even with existing committees, there are certain topics that need to be highlighted, to ensure that the industry is paying sufficient importance to them.

SAE EDGE™ reports are intended to be quick, concise overviews of major unsettled areas where vital new technologies are emerging. By definition, an unsettled area is characterized more by confusion and controversy than established order. Early practitioners must confront an absence of agreement. Their challenge is often not to seize the high ground but to find common ground. These scouting reports from the frontiers of investigation are intended merely to begin the process of sorting through critical issues, contributing to a better understanding of key problems, and providing helpful suggestions about possible next steps and avenues of investigation.

SAE EDGE™ Research Reports, therefore, are fundamentally distinct from the more formal working groups approach and far removed from the more mature cooperative research project and standards development process. They begin by convening an informal meeting of, or extensively interviewing, subject matter experts working in an unsettled area of technology from industry, government, and the research community. A moderated but collegial discussion ensues regarding their shared interest. This conversation attempts to identify critical issues hindering a clear understanding of important topics, preventing a common agreement on potential solutions, or slowing progress in key areas of research.

An SAE EDGE™ Research Report is then drafted and circulated for comment to meeting attendees and other interested parties. After review, the final result of this process is a report like the current examination of unsettled issues concerning electric propulsion for aircraft.

## Abbreviations/Definitions

**AE** - All-Electric

**BMS** - Battery Management System

**CTO** - Chief Technology Officer

**DARPA** - Defense Advanced Research Projects Agency

**eVTOL** - Electric Vertical Takeoff and Landing

**GA** - General Aviation

**GHG** - Greenhouse Gas

**GRC** - Glenn Research Center (NASA)

**GT** - Gas Turbine

**HE** - Hybrid Electric

**HEP** - Hybrid Electric-Parallel

**HES** - Hybrid Electric-Series

**IEC** - International Electrotechnical Commission

**ME** - More-Electric

**NASA** - National Aeronautics and Space Administration

**NEAT** - NASA Electric Aircraft Testbed

**OEM** - Original Equipment Manufacturer

**PEHV** - Plug-in Electric-hybrid Vehicle

**PE** - Power Electronics

**UAM** - Urban Air Mobility

## Acknowledgments

While I am responsible for putting this report together, I got a lot of technical input from some of the most experienced aerospace experts. I acknowledge them alphabetically, starting with Richard Ambroise, who is the Head of Propelling for the Electrification Flight Demonstrator program at Airbus UpNext which is like a startup within the larger OEM. He is also the founding chair of SAE's E-40 technical committee on Electrified Propulsion. Rodger Dyson is a leading researcher at NASA Glenn, who is also the Principal Investigator of the NASA Electric Aircraft Testbed (NEAT) facility where NASA is testing out different power systems for electric aircraft at all power levels. Craig Lawson is a senior lecturer at Cranfield University, who has been a source of great information about the system level processes associated with electric aircraft. He is one of the creators of a popular graduate level program in the Centre for Aeronautics on aircraft design. Steve O'Flarity is the Associate Director, Systems Analysis and Validation, at Pratt & Whitney and is involved in various technical areas including electric propulsion. He was one of the leaders of the team that developed engine health monitoring within P&W and served for a while as the Executive Technical Assistant to the Vice President – Engineering Technology and Environment. Norman Pereira is an Aerospace Engineer and a Verified AHR-Manager within FAA's Electrical & Mechanical Equipment Section. He is an expert on battery systems and has worked with the RTCA and SAE International in developing standards for aviation batteries.

Joshua Portlock is the CTO and Co-Founder of a startup in Australia - Electro.Aero Pty Ltd, that develops electric propulsion systems, including chargers, for next generation light aircraft. Gilles Rosenberger cofounded Faraday Aerospace near Paris after working for many years in the aerospace industry in leadership positions at Safran and Airbus (significantly as the chief operating officer of VoltAir).

Michael Winter is the senior fellow for advanced technology at Pratt & Whitney, United Technologies Corporation, who has held many leadership roles there and at United Technologies Research Center. Without their input and ready accessibility, this report would not have been written or would have been much less informative. Finally, I must thank Monica Nogueira from SAE, who was the impetus behind this work and her constant "encouragement" made sure that it was completed in time for publication!

## References


1. Aerospace Technology Institute, "Electrical Power Systems," Insight, July 2018.
2. AIN, "Eviation Completes Prototype of Alice Electric Aircraft," June 2019, https://www.ainonline.com/aviation-news/air-transport/2019-06-16/eviation-completes-prototype-alice-electric-aircraft. Retrieved October 2019.
3. Air Transport Action Group, "Aviation's Impact on the Environment," https://aviationbenefits.org/environmental-efficiency/aviations-impact-on-the-environment/. Retrieved October 2019.
4. Ball, J., "The Race Is on to Build a Better Battery," Forbes, May 2019.
5. Bogaisky, J., "Boeing Backs away from Zunum Aero, Founders Struggle to Raise Money after Laying Off Staff," Forbes Aerospace and Defense, July 2019, https://www.forbes.com/sites/jeremybogaisky/2019/07/02/zunum-aero-nears-zero-founders-struggle-to-raise-money-after-laying-off-staff/#3f6c955d2d73. Retrieved October 2019.
6. Dyson, R.W., "NASA Electric Aircraft Testbed (NEAT) Single-Aisle Transport Air Vehicle Hybrid Electric Tail-Cone Thruster Powertrain Configuration and Test Results," in *Proceedings of the AIAA Propulsion and Energy Forum, 2018 AIAA/IEEE Electric Aircraft Technologies Symposium*, Cleveland, OH, July 2018.
7. Dyson, R., Jansen, R., Duffy, K., and Passe, P., "High-Efficiency Megawatt Machine Rotating Cryocooler Conceptual Design," in *AIAA/IEEE Electric Aircraft Technology Symposium*, Indianapolis, IN, August 2019.
8. Epstein, A.H. and O'Flarity, S.M., "Considerations for Reducing Aviation's CO2 with Aircraft Electric Propulsion," *Journal of Propulsion & Power* 35(3):572-582, May-June 2019.
9. eVTOL news, "Kitty Hawk Cora," https://evtol.news/aircraft/kitty-hawk-cora/. Retrieved October 2019.
10. Fehrm, B., "Bjorn's Corner: Electric Aircraft, the First Fall on the Hype Curve," https://leehamnews.com/2019/05/31/bjorns-corner-electric-aircraft-the-first-fall-on-the-hype-curve/. Retrieved September 2019.
11. Hartwig, J., Fraser, B., Brown, G., Koci, D., Hunker, K. et al., "A New Liquid Hydrogen Based Superconducting Coil Test Rig to Measure AC Losses," in *Proceedings of the AIAA Propulsion and Energy Forum, 2018 AIAA/IEEE Electric Aircraft Technologies Symposium*, Cleveland, OH, July 2018.
12. ICAO, "Aircraft Engine Emissions," https://www.icao.int/environmental-protection/Pages/aircraft-engine-emissions.aspx. Retrieved September 2019.
13. IEC, "Electric Vehicle Conductive Charging System," IEC 61851-1:2017-Several Parts, February 2017.
14. Jansen, R., Madavan, N., and Dyson, R., *Case for Electrifying Large Electric Aircraft*, Book chapter advanced copy (Reston, VA: AIAA Press, 2020).
15. Koehler, T., *Being makes history with flights of Fuel Cell Demonstrator Airplane* (Chicago, IL: Boeing Frontiers, May 2008).
16. Koff, B.L., "Spanning the Globe with Jet Propulsion," in *AIAA 21st Annual Meeting and Exhibit*, Arlington, VA, April-May, 1991.
17. Momar, D.S. and Evans, D.D., "Major U.S. Cities Using National Standard Fire Hydrants, One Century after the Great Baltimore Fire," NISTIR 7158, NIST, August 2004.
18. Portlock, J., "International Charging Standard for Electric Aircraft and Modular Electric Propulsion Systems," Presented at in *SAE International EASG Meeting*, Oshkosh, WI, July 2019.
19. Rajamani, R., *Electric Flight Technology: The Unfolding of a New Future* (Warrendale, PA: SAE International, April 2018).

20. Renewable Energy Magazine, "Ampaire Preparing to Test Electric EEL in Partnership with Mokulele Airlines," August 2019, https://www.renewableenergymagazine.com/electric_hybrid_vehicles/ampaire-preparing-to-test-electric-eel-in-20190816. Retrieved October 2019.
21. Rheaume, J. and Lents, C., "Energy Storage for Commercial Hybrid Electric Aircraft," SAE Technical Paper 2016-01-2014, 2016, https://doi.org/10.4271/2016-01-2014.
22. RTCA, "Minimum Operational Performance Standards for Rechargeable Lithium Batteries and Battery Systems," DO-311, December 2017.
23. Stenquiest, P., "How Green Are Electric Cars? Depends on Where You Plug In," *New York Times*, April 13, 2012.
24. SAE International, "SAE Electric Vehicle and Plug in Hybrid Electric Vehicle Conductive Charge Coupler," J1772, 2017.
25. SAE International, "SAE Energy Transfer System for Electric Vehicle," J2293 (Parts 1 and 2), 2017
26. Thalin, P., editor, *Fundamentals of Electric Aircraft* (Warrendale, PA: SAE International, December 2018).
27. Thompson, R., "Global Electric Aircraft Developments Continue Rapid Upward Trend," Roland Berger report, May 2019, https://www.rolandberger.com/en/Point-of-View/Electric-propulsion-is-finally-on-the-map.html. Retrieves October 2019.
28. Waller, M., *The Electrification of Civil Aircraft and the Evolution of Energy Storage* (Warrendale, PA: SAE International, August 2017).
29. Tilly, E.A., artist, Airship Powered by an Electric Motor Developed by Albert and Gaston Tissandier Departing from Auteuil, Paris, France, October 8, 1883]/Poyet, del ; E. A. Tilly, 1883.
30. Balachandran, T., Lee, D., Salk, N., and Haran, K.S., "A Fully Superconducting Air-Core Machine for Aircraft Propulsion," in *CEC-ICMC Conference*, Hartford, CT, July 21-25, 2019.

## Contact Information

EDGEresearch@sae.org.

The authors of this document together with the SAE Team responsible for its creation join in expressing their deepest appreciation to all of the individuals who contributed.

www.ingramcontent.com/pod-product-compliance
Lightning Source LLC
LaVergne TN
LVHW070937210926
848482LV00003B/268
*9781468601152*